DWA

DIGITAL WEALTH

ACTIVATION

A Step-by-Step Roadmap to Building Profitable Online Businesses Through Drop shipping & E-Commerce

Rachell Jovan and Steve Barns

Table of Contents

Preface

The digital economy is evolving at lightning speed, and with it comes endless opportunities for anyone willing to learn, adapt, and act. This book is born out of the need to demystify online business—breaking it down into simple, actionable steps that anyone, regardless of background, can follow. Whether you're a beginner curious about online income or an aspiring entrepreneur ready to scale, this book will serve as your companion and guide.

The goal is not just to give you information, but to empower you with the mindset, tools, and strategies to thrive in the online world. By the time you reach the last page, you won't just know what to do—you'll feel confident enough to start.

Why This Book Matters

So much advice online is either too vague to act on, or so technical that it feels overwhelming. This book bridges that gap. It provides clarity, structure, and inspiration. More importantly, it lays out **practical roadmaps** for building and sustaining an online business.

The world is shifting to digital-first living—shopping, learning, working, and connecting online. Those who understand how to create value in this space will always remain ahead. This book is about giving you that edge.

How to Use This Book

Think of this book as both a **manual** and a **mentor**. Each chapter builds on the last, guiding you step by step through key concepts, business models, and success strategies.

Here's how you can get the most out of it:

- **Read actively**: Highlight, underline, or jot down notes.
- **Apply as you go**: Don't just read—try the exercises, explore the tools, and test out the strategies.
- **Adapt to your context**: Every reader's journey is unique. Take what works for you and customize it.
- **Revisit often**: Online business changes fast. Use this book as a reference guide to refresh your strategies.

If you follow along with openness and commitment, you'll finish with not just knowledge—but with a plan you can act on immediately.

About the Author

Rachell Jovan and Steve Barns are passionate entrepreneurs, digital strategists, and mentors dedicated to helping individuals unlock the opportunities of the online world. With years of experience navigating the complexities of e-commerce, content creation, and digital marketing, she has built a reputation for breaking down complex ideas into simple, actionable steps.

Their mission is to empower everyday people—students, professionals, and aspiring entrepreneurs alike—to build profitable, sustainable online businesses. Beyond their professional pursuits, Rachell and Steve believes in continuous learning, resilience, and the power of taking bold, consistent action.

This book reflects their journey, insights, and commitment to guiding others on the path of online business mastery.

Part I – Foundations of Online Business

"Don't build a mansion on sand. Every great online business rests on solid foundations."

Before diving into strategies, tools, and income models, you must understand the **building blocks of online business success**. Without these foundations—choosing the right model, selecting a profitable niche, and positioning your brand—you'll struggle with inconsistency and frustration.

This part lays the groundwork so you start on the right foot.

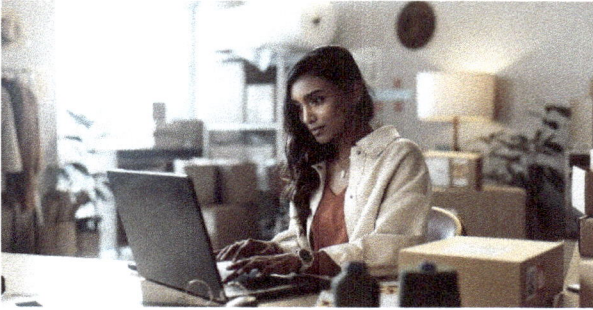

Chapter 1: Choosing Your Online Business Model

Every successful online entrepreneur begins with one critical decision: selecting the right business model. Just like an architect chooses the foundation before constructing a skyscraper, your business model forms the bedrock of everything you will build. There is no "one-size-fits-all" here—your choice must align with your strengths, resources, long-term vision, and tolerance for risk.

In this chapter, you'll explore the most common online business models—affiliate marketing, blogging, e-commerce, digital products, and more. You'll learn how each works, the startup requirements, revenue potential, and the lifestyle it creates. For example, affiliate marketing may suit someone who enjoys content creation but wants to avoid inventory headaches, while e-commerce works best for entrepreneurs who want full control over physical or digital product sales.

By the end of this chapter, you will understand that choosing your business model is not about chasing trends but about aligning your personal goals with a sustainable system. Think of it as planting the right seed in fertile soil—you'll water it, nurture it, and watch it grow into your business empire.

The internet has democratized business. Today, anyone with a smartphone and determination can start an online venture. But the first question is:

Which model is right for me?

Here are four of the most popular paths:

1. Affiliate Marketing

You promote other people's products through links and earn a commission per sale.

- **Real Example:** Pat Flynn of *Smart Passive Income* famously built his empire through affiliate marketing, recommending software, courses, and tools he genuinely used.
- **Pros:** Low start-up costs, flexible, scalable.
- **Cons:** You rely on someone else's product quality and policies.

2. Blogging for Income

You create a blog, build an audience, and monetize through ads, affiliate links, digital products, or sponsorships.

- **Real Example:** Michelle Schroeder-Gardner (*Making Sense of Cents*) earns over $100,000 per month through blogging, primarily via affiliate programs.
- **Pros:** Builds authority, multiple income streams.
- **Cons:** Takes time to grow traffic and credibility.

3. Drop shipping (E-commerce)

You sell physical products online without holding inventory. Suppliers ship directly to your customers.

- **Real Example:** Gymshark, now worth over a billion dollars, started small by leveraging e-commerce and outsourcing production.
- **Pros:** No warehouse needed, potentially high scale.
- **Cons:** Thin margins, customer service challenges, competition.

4. Digital Business Automation

You set up automated systems—email sequences, sales funnels, chatbots, webinars—that generate income while you focus on strategy.

- **Real Example:** Russell Brunson's *ClickFunnels* is built around the power of automation.
- **Pros:** Scalable, time-saving, creates passive income systems.
- **Cons:** Requires technical setup and consistent optimization.

Checklist: Choosing Your Business Model

☑ Do I want to sell my own product, or promote others'?

☑ Do I prefer writing/content creation, or selling physical products?

☑ Do I want something quick and flexible (affiliate) or long-term authority (blogging)?

☑ Am I willing to learn tech tools for automation?

☑ Which model excites me most?

Action Step: Circle the one model that best aligns with your skills and interests. If unsure, pick one to experiment with for the next 90 days.

Chapter 2: Niche Selection & Validation

If choosing your business model is the "what," then selecting your niche is the "where." A niche defines the market you'll serve, the audience you'll attract, and the problems you'll solve. Many beginners fail not because they lack passion or effort, but because they pick the wrong niche—either too broad, too competitive, or too unprofitable.

In this chapter, you'll learn how to identify profitable niches, test market demand, and validate your ideas before committing time and money. Validation ensures that there's both an audience who needs your solution and a willingness to pay for it. You'll discover practical methods like keyword research, trend analysis, competitor studies, and pre-launch surveys.

The goal is to strike a balance between passion, profitability, and market demand. A niche you love but no one cares about won't pay the bills. A profitable niche you hate will burn you out. But a niche that excites you, has growing demand, and solves real problems can become a goldmine. By the end of this chapter, you'll be equipped to say with confidence: *"Yes, this is the niche where I will build my empire."*

You can't sell to "everyone." A profitable online business starts with targeting a **niche**—a specific group of people with a shared problem, passion, or desire.

Think of niches like *fishing ponds*. Some ponds are overfished (too competitive), while others are too small (no buyers). The key is finding one with **enough hungry fish** that matches what you can provide.

The Success Formula: Passion + Demand + Profitability

1. **Passion** – Do you genuinely enjoy the topic? This keeps you motivated long-term.
2. **Demand** – Are people searching for solutions? (Check Google Trends, Amazon, Reddit, Quora.)
3. **Profitability** – Are people spending money here? Look for existing products, ads, or affiliate programs.

Example Niches

- **Health & Fitness:** weight loss, yoga, meal planning.
- **Personal Finance:** budgeting, investing, side hustles.
- **Lifestyle:** minimalism, parenting hacks, productivity.
- **Hobbies:** gaming, photography, gardening.

Case Study: The Fitness Blogger

Sarah started a blog about general wellness. She struggled until she niched down to "yoga for busy moms." That narrow focus attracted loyal readers and high-paying affiliate opportunities in yoga mats, online classes, and wellness products.

Checklist: Validating Your Niche

☑ Use Google Trends to confirm people are searching.

☑ Search Amazon: Are there bestsellers in this category?

☑ Look at competitors: Are they making money?

☑ Test engagement: Post in a Facebook group or create a small ad.

Action Step: Write down 3 niche ideas. Run them through the checklist. Pick the one with the strongest *passion + demand + profitability*.

Chapter 3: Branding and Positioning for Success

In the noisy digital world, your brand is not just your logo, colors, or slogan—it's the story you tell, the promise you make, and the emotional connection you create with your audience. Whether you're a solo blogger, a dropshipper, or an e-commerce mogul, branding is what separates you from being "just another seller" to becoming a trusted authority.

This chapter explores the psychology of branding and how to position yourself effectively in a crowded marketplace. You'll learn how to craft a unique value proposition (UVP), build consistency across platforms, and create a voice that resonates with your target audience. We'll also cover the importance of trust-building elements like social proof, customer reviews, storytelling, and visual identity.

Positioning, on the other hand, ensures you occupy the right space in your customer's mind. Are you the affordable option? The premium expert? The niche authority? This chapter will help you decide how to stand out and be remembered.

At the end of this chapter, you'll understand that branding is not a one-time task but an ongoing process. It's about being intentional in how you show up, ensuring that every touchpoint—from your website to your social media—communicates reliability, relatability, and relevance.

Branding isn't just a logo. It's the **feeling your audience gets when they interact with you**. Strong branding builds trust, recognition, and loyalty.

The 4 Pillars of a Strong Brand

1. **Clarity** – Be clear about who you serve.

   Example: "I help young professionals save money and escape debt."

2. **Consistency** – Use the same tone, colors, and visuals everywhere.

3. **Trust** – Share authentic stories, testimonials, and valuable content.

4. **Positioning** – Decide: Are you the affordable choice, luxury expert, or niche specialist?

Case Study: Dropshipping Store Positioning

Two stores sell water bottles. One brands itself as "cheap everyday bottles." The other positions as "eco-friendly, sustainable bottles for health-conscious millennials." Guess which one earns more loyal customers and can charge higher prices? (The second.)

Checklist: Branding Foundations

☑ Write your **brand statement**: *I help [who] achieve [result] through [method].*

☑ Choose 2–3 brand colors and stick with them.

☑ Decide your tone: professional, friendly, or casual.

☑ Create a simple logo (using Canva or Fiverr).

☑ Be consistent across your website, social media, and emails.

Action Step: Draft your one-line brand statement today. Example: *I help busy moms stay fit at home through simple yoga routines.*

Closing for Part I

You now have the **foundations**: a chosen business model, a validated niche, and the beginnings of your brand. Think of this part as laying down the **blueprint** of your online business house. Without it, all the fancy marketing and automation in later chapters won't hold.

As we move into **Part II – Affiliate Marketing Mastery**, we'll explore how to turn these foundations into real income by promoting products online.

Key Takeaways from Part I:

- Pick one business model and commit to testing it.
- Validate your niche with passion, demand, and profitability.
- Build a trustworthy brand that positions you clearly in the market.

Part II – Affiliate Marketing Mastery

"Affiliate marketing has made businesses millions and ordinary people millionaires." – Bo Bennett

Affiliate marketing is one of the fastest ways to start making money online. You don't need to create a product, handle shipping, or deal with customer service. Instead, you connect people with products they already want and get rewarded with a commission.

But success doesn't come from spamming links everywhere. It requires **strategy, systems, and the right mindset**. In this section, you'll learn how to build an affiliate business that is ethical, sustainable, and highly profitable.

Chapter 1: Understanding the Affiliate Ecosystem (Main Players)

B efore diving headfirst into affiliate marketing, you need to understand the ecosystem that drives it. Affiliate marketing isn't just about promoting a product and earning a commission—it's a carefully structured system with several key players who make it all work. At the center is **you, the affiliate marketer**, but your role is tied to merchants (the product creators or service providers), affiliate networks (the middlemen connecting affiliates with merchants), and of course, the customers whose trust you must earn.

This chapter introduces you to each of these players and breaks down how money flows through the ecosystem. You'll learn how Amazon Associates, ClickBank, and CJ Affiliate function differently, and why understanding commission structures, cookie durations, and payout terms can mean the difference between profit and loss. By the end of this chapter, you'll see affiliate marketing as an interconnected web of relationships, not just a one-off sales tactic.

Affiliate marketing works like a triangle:

1. **The Merchant/Advertiser** – The company that owns the product (e.g., Amazon, Bluehost, ClickBank vendors).
2. **The Affiliate (You)** – The promoter who markets the product and earns a commission.
3. **The Customer** – The buyer who purchases through your link.

Sometimes, a fourth party is involved:

- **Affiliate Networks** (like CJ Affiliate, ShareASale, or ClickBank), which act as middlemen connecting you with multiple merchants.

Example:

- You sign up for the Amazon Associates program.
- You write a blog post reviewing "best fitness trackers for 2025."
- A reader clicks your link and buys a $150 tracker.
- You earn a commission (typically 4–8%).

Key Insight: Affiliates who provide **real value** (like reviews, comparisons, or tutorials) earn much more than those who just drop links.

Chapter 2: Choosing Products & Niches

Success in affiliate marketing hinges on **choosing the right niche and products**. Too broad, and you'll drown in competition. Too narrow, and you'll run out of audience or opportunities. This chapter equips you with strategies to find the sweet spot: a niche you're passionate about, profitable enough to sustain growth, and backed by real consumer demand.

You will learn the importance of validating niches with tools like Google Trends, SEMrush, or even simple Amazon best-seller lists. This chapter also explores **product selection**—why it's crucial to promote items you either use yourself or that align with your audience's problems and aspirations. Whether it's software, courses, physical products, or subscriptions, the products you choose define your brand identity and determine how your audience perceives your credibility. By the end, you'll know how to avoid "shiny object syndrome" and focus on niches with both **longevity and profitability.**

Your success hinges on promoting the **right products to the right people**.

High-Ticket vs Low-Ticket

- **Low-Ticket:** Cheap, everyday products (like books, clothing). Easier to sell, but you need a lot of volume.
- **High-Ticket:** Expensive items (like online courses, software, or premium gear). Fewer sales needed, but requires more trust-building.

Case Study:

- John started promoting $20 phone accessories on Amazon. After months, his income was inconsistent.
- He switched to promoting $500 digital cameras and $1,000 online courses in photography. His income multiplied, even though he sold fewer units.

Checklist: Choosing Affiliate Products

☑ Does it solve a real problem?

☑ Would I personally use or recommend it?

☑ Is the commission rate worth it?

☑ Are there recurring commissions (like software subscriptions)?

☑ Is the company reputable?

Action Step: Choose 2–3 products you'd be comfortable recommending to a friend.

Chapter 3: Promoting Affiliate Offers (SEO, Blogging, YouTube, Social Media)

Finding good products is just the beginning—promotion is where you turn opportunities into income. This chapter introduces the **four powerhouse promotional channels** every affiliate should consider: **SEO, blogging, YouTube, and social media.**

Here, you'll learn why SEO is the long game that compounds over time, how blogging builds authority and trust, why YouTube is the perfect mix of storytelling and education, and how social media creates fast visibility and community engagement. Each method has its strengths—

SEO might bring steady organic traffic, while TikTok or Instagram can provide viral bursts of attention.

This chapter also compares free vs. paid promotion, showing you when it's wise to invest in ads and when it's better to double down on organic strategies. The introduction frames promotion not as "spamming links" but as **delivering genuine value**, solving problems, and positioning yourself as a trusted guide.

Promotion is where most affiliates fail. Dropping links without context doesn't work. Instead, think of affiliate marketing as **helping your audience make better buying decisions**.

Popular Promotion Channels:

1. **Blogging & SEO** – Write product reviews, comparisons, and tutorials.
   - Example: "Best Laptops for College Students (2025 Guide)."
2. **YouTube** – Create product reviews, unboxing videos, and tutorials.
   - Example: "How to Build a Website in 10 Minutes (Bluehost Tutorial)."
3. **Email Marketing** – Build a list and send helpful tips, with affiliate links inside.
4. **Social Media** – Instagram, TikTok, Pinterest for product showcases.
5. **Webinars** – Great for high-ticket courses or software.

Pro Tip: Always disclose your affiliate links. Transparency builds trust.

Chapter 4: Email Marketing & Lead Magnets for Affliates

In affiliate marketing, **the money is in the list.** While social platforms can disappear overnight, an email list is an asset you own—a direct line to your audience's inbox. This chapter introduces you to the world of email marketing and explains why it's the lifeblood of sustainable affiliate income.

The introductory section emphasizes **relationship-building over hard selling.** Instead of bombarding subscribers with links, you'll learn the art of nurturing leads, delivering value, and using storytelling to build trust.

This chapter also explores **lead magnets**—freebies like eBooks, checklists, or mini-courses that attract subscribers in exchange for their email addresses. By the end, you'll see email not just as a promotional channel but as a long-term trust and revenue machine that can generate consistent sales on autopilot.

If you rely only on one-time clicks, you'll struggle. The real money is in building an **email list**—a group of people you can market to repeatedly.

How to Do It:

1. Create a **lead magnet** (e.g., free guide, checklist, or mini-course).
2. Use a landing page to collect emails.
3. Send a **welcome sequence** introducing yourself and offering value.
4. Recommend affiliate products naturally within your emails.

Example: A personal finance blogger gives away a free "Budget Template." In the follow-up emails, she recommends budgeting software with her affiliate link.

Chapter 5: High Ticket vs Low Ticket Strategies

Affiliate marketing thrives on commissions — but not all commissions are created equal. Some affiliates focus on **low-ticket products** (small commissions, but higher sales volume), while others go after **high-ticket offers** (larger commissions per sale, but fewer customers). Both models can work, but success depends on how you structure your strategy, audience, and marketing approach.

This chapter will give you a **deep dive into both worlds**, helping you understand how to choose, promote, and profit from either (or both).

1. What Are Low-Ticket Affiliate Products?

- **Definition:** Products priced between $5 – $100 (sometimes up to $300).
- **Commission Range:** Usually $2 – $50 per sale.
- **Examples:**
  - Amazon products (books, gadgets, kitchenware).
  - Online subscriptions (Spotify, Canva, Grammarly).
  - Entry-level digital courses or eBooks.

Advantages of Low Ticket Products

- Easier to sell (low price = less customer hesitation).
- Higher conversion rates with impulse buyers.
- Great for beginners learning marketing skills.

Disadvantages

- Requires high volume to make meaningful income.
- Marketing costs (ads, content creation) can outweigh small commissions.

2. What Are High-Ticket Affiliate Products?

- **Definition:** Products priced at $500 – $10,000+.
- **Commission Range:** $100 – $5,000 per sale.
- **Examples:**

    - Premium software (ClickFunnels, Kajabi, SEMrush).
    - High-end coaching/consulting programs.
    - Luxury goods (watches, travel packages).
    - Expensive online courses/masterminds.

Advantages of High Ticket Products

- Fewer sales needed to hit income goals.
- Can justify spending more on ads/content since margins are higher.
- Attracts a serious, invested customer base.

Disadvantages

- Harder to sell — customers need trust, proof, and nurturing.
- Longer sales cycle (people don't spend $2,000 in one click).
- Requires strong branding and authority.

3. Comparing the Two Models

Feature	Low-Ticket	High-Ticket
Price Range	$5 – $300	$500 – $10,000+
Commission	$2 – $50	$100 – $5,000
Sales Volume Needed	Very high	Low
Sales Cycle	Short (impulse)	Long (trust-building)
Beginner Friendly?	Yes	Harder
Scalability	Limited without ads	More scalable

4. Choosing the Right Strategy

Your choice depends on:

- **Audience Size** – Small audiences may benefit from high-ticket (since you don't need thousands of buyers).
- **Trust & Authority** – If you're new, low-ticket may work better at first.
- **Marketing Channels** – Social media & blogs favor low-ticket; webinars & email funnels favor high-ticket.
- **Goals** – Do you want quick wins (low-ticket) or long-term, fewer but bigger payouts (high-ticket)?

5. Marketing Strategies for Low-Ticket Offers

1. **Volume-Based Content Creation**
   - Write lots of blogs, reviews, YouTube videos.
   - Use SEO to attract large numbers of people.
2. **Impulse Marketing**
   - Highlight convenience, speed, and affordability.
   - Example: "Best Budget Microphones for YouTube Under $50."
3. **Bundles & Lists**
   - Promote "Top 10 Must-Have Tools for Students" instead of one item.
4. **Social Media Influence**
   - TikTok, Instagram, and Pinterest work well for quick, cheap product showcases.

6. Marketing Strategies for High-Ticket Offers

1. **Authority Building**
   - o Publish in-depth guides, case studies, and thought leadership content.
   - o Example: A 5,000-word blog post on "How to Start a 7-Figure Coaching Business."
2. **Webinars & Workshops**
   - o Run live or recorded webinars that educate and pitch the high-ticket solution.
3. **Email Sequences & Nurturing**
   - o Long-term engagement builds trust before asking for a big purchase.
4. **Storytelling & Testimonials**
   - o Share real-life success stories to show transformation.
5. **One-on-One Consulting Before Sale**
   - o Many high-ticket programs require calls or DMs before closing.

7. Hybrid Strategy (Best of Both Worlds)

Many successful affiliates combine **low-ticket + high-ticket** offers in one funnel:

- Step 1: Attract leads with **low-ticket or free offers** (ebooks, $20 software, $50 gadgets).
- Step 2: Nurture them via email with valuable tips.
- Step 3: Upsell them to **high-ticket programs** once they trust you.

Example Funnel:

- Free guide → Affiliate link for $29 tool → Email nurturing → $997 online course.

8. Case Studies

Case Study 1 – Low Ticket Success

A fitness blogger reviews affordable gym equipment and earns **$2,500/month** through Amazon Associates. Although commissions are tiny, the huge traffic from SEO (200k visitors/month) makes it work.

Case Study 2 – High Ticket Success

A digital marketing coach runs a webinar funnel for a $2,000 course. With only **100 monthly attendees**, converting 5% gives **$10,000/month** in commissions.

9. Common Mistakes to Avoid

- **Low Ticket Mistakes:** Relying on ads (not profitable), ignoring SEO, promoting random low-value products.
- **High Ticket Mistakes:** Skipping trust-building, pushing too hard, promoting scams or overpriced programs.

10. Best Practices

- Start with **low-ticket** to build skills & audience.
- Transition to **high-ticket** once you've built trust.
- Always **mix in email marketing** — it's the bridge between both worlds.
- Think in terms of **lifetime customer value**, not single sales.

Chapter 6: Common Mistakes & Myths in Affiliate Marketing

Affiliate marketing is filled with opportunities, but it's also littered with pitfalls. Shiny offers, fraudulent networks, spammy tactics, and unrealistic expectations have trapped countless beginners. This chapter introduces you to the **red flags and mistakes** that can ruin your credibility, finances, or both.

The opening emphasizes that trust is your greatest currency as an affiliate. Promoting products you don't believe in, stuffing links everywhere, or falling for "get rich quick" schemes can burn your audience and derail your progress. Here, you'll be guided on how to **vet affiliate programs**, recognize predatory scams, and sidestep common errors like over-relying on one traffic source.

This chapter sets the stage for resilience—teaching you that success in affiliate marketing is built on **patience, ethical choices, and sustainable strategies.**

Mistakes to Avoid

✗ Promoting too many random products.

✗ Copy-pasting product descriptions (Google will ignore your content).

✗ Hiding affiliate links (kills trust).

✗ Expecting instant results.

Myths

- *"Affiliate marketing is a get-rich-quick scheme."*
- *"You need a massive audience to make money."*
- *"Only tech-savvy people succeed."*

Truth: With patience, strategy, and the right niche, affiliate marketing is accessible to anyone.

Chapter 7: Best Practices & Case Studies in Affiliate Marketing

B y now, you've explored the foundations of affiliate marketing—how to pick a niche, promote offers, and structure campaigns. But knowing the "what" is only half the battle. The "how" is where many aspiring affiliates either break through or burn out. That's why learning **est practices** and studying **real-life case studies** is critical.

This chapter gives you the playbook: the strategies top affiliates consistently follow, mistakes to avoid, and real examples of how ordinary people turned affiliate marketing into six- and seven-figure businesses. By combining proven best practices with inspirational case studies, you'll not only learn what works—you'll believe that you can do it too.

Section 1: Best Practices in Affiliate Marketing

1. Focus on Building Trust, Not Just Sales

- **Why it matters**: Affiliate marketing thrives on trust. If your audience feels you only push products for commission, they'll leave. If they trust your recommendations, they'll buy again and again.
- **How to apply**: Share authentic experiences, include pros and cons in reviews, and only promote products you'd recommend to a friend.

2. Niche Down and Stay Consistent

- **Why it matters**: Broad niches confuse your audience and dilute your authority. Narrow niches make you the "go-to" person.
- **How to apply**: Instead of "health," focus on "keto diets for busy professionals" or "yoga for seniors."

3. Leverage Multiple Traffic Sources

- **Why it matters**: Depending on one platform is risky. A YouTube ban, Google update, or TikTok algorithm change can wipe out your business overnight.
- **How to apply**: Diversify with SEO, YouTube, email marketing, and social media. Think long-term.

4. Master Email Marketing Early

- **Why it matters**: Email converts better than any other traffic source because it allows personal, repeated contact with your audience.

- **How to apply**: Use lead magnets (ebooks, free courses, templates), set up automated sequences, and nurture your subscribers with value-first content.

5. Prioritize Evergreen Content

- **Why it matters**: Evergreen content (that remains relevant over time) keeps generating traffic and commissions long after you create it.
- **How to apply**: Blog posts like "Best Tools for Freelancers 2025" or "Beginner's Guide to Keto" stay useful year after year, unlike seasonal promotions.

6. Track and Optimize

- **Why it matters**: Data-driven affiliates outperform guessers. If you don't know which ad, keyword, or funnel step converts, you can't scale.
- **How to apply**: Use tracking tools (Google Analytics, Voluum, ClickMagick). Test headlines, CTAs, and landing pages continuously.

7. Invest in Skills, Not Just Tools

- **Why it matters**: Tools can make you efficient, but skills make you profitable. Skills like copywriting, SEO, and storytelling have a lifetime ROI.
- **How to apply**: Dedicate time each week to learning and improving instead of chasing shiny new tools.

8. Think Long-Term, Not Just Quick Wins

- **Why it matters**: Many affiliates chase fast commissions and burn out. Those who focus on building a brand and email list build businesses that last.
- **How to apply**: Create content and strategies that will still bring in revenue years from now.

Section 2: Case Studies of Successful Affiliate Marketers

Nothing beats real-world inspiration. Let's look at some affiliates who started small, applied best practices, and scaled massively.

Case Study 1: Pat Flynn – Smart Passive Income

- **Niche**: Online business & passive income.
- **Journey**: Laid off from his job as an architect in 2008. Started blogging to share exam notes, then moved into affiliate marketing.
- **Best Practices Applied**:
    - Built deep trust with his audience by being transparent about his income reports.

- o Created evergreen content that ranked in search engines.
- o Focused on email marketing early.
- **Result**: Generates **$100,000+ monthly**, primarily from affiliate partnerships with hosting companies, software tools, and courses.

Case Study 2: Michelle Schroeder-Gardner – Making Sense of Cents

- **Niche**: Personal finance & lifestyle.
- **Journey**: Started her blog as a side hustle while working as a financial analyst. Began promoting budgeting and finance tools.
- **Best Practices Applied**:
  - o Niched down to a profitable, high-demand space (personal finance).
  - o Leveraged SEO and Pinterest for consistent traffic.
  - o Built trust by sharing her debt-free journey.
- **Result**: Earned over **$1.5 million annually**, much of it from affiliate programs like financial tools and budgeting apps.

Case Study 3: Wirecutter (Acquired by New York Times)

- **Niche**: Product reviews & recommendations.
- **Journey**: Started as a small site reviewing gadgets and electronics. Focused on detailed, honest, and SEO-friendly reviews.
- **Best Practices Applied**:
  - Prioritized evergreen content.
  - Invested heavily in SEO and user experience.
  - Built authority by publishing detailed, data-backed reviews.
- **Result**: Scaled so successfully that the **New York Times acquired it for $30 million** in 2016.

Case Study 4: Spencer Haws – Niche Pursuits

- **Niche**: SEO, online business, software tools.
- **Journey**: Experimented with multiple niche websites before creating Niche Pursuits and the keyword research tool "Long Tail Pro."
- **Best Practices Applied**:
  - Focused on evergreen content and SEO traffic.
  - Promoted his own software alongside affiliate offers.
  - Shared his wins and failures transparently to build trust.
- **Result**: Runs a six-figure affiliate and software business.

Case Study 5: Unknown Micro-Niche Blogger (Fitness Example)

- **Niche**: "Keto recipes for beginners."
- **Journey**: Started a blog in 2020 during the pandemic. Created step-by-step guides, recipe posts, and YouTube cooking tutorials.
- **Best Practices Applied**:
  - Very narrow niche selection.
  - Built an email list with free "7-Day Keto Meal Plan."
  - Used affiliate links for keto cookbooks, supplements, and meal plans.
- **Result**: Within 18 months, earning **$7,000/month** from affiliate sales. Proof that even new bloggers can succeed when consistent.

Section 3: Lessons Learned

From these best practices and case studies, several universal lessons emerge:

1. **Authenticity beats hype.** The most successful affiliates are those who genuinely help their audience.
2. **Traffic is king, but trust is the crown.** Without audience trust, no amount of traffic converts.
3. **Evergreen + consistent = compound growth.** Content works for you long after it's published.
4. **Diversify for stability.** Don't depend on one traffic source or one affiliate partner.
5. **Patience pays.** Affiliate marketing is not a get-rich-quick scheme. Success is built over time.

Conclusion

Affiliate marketing success is not reserved for a lucky few—it's available to anyone who is consistent, strategic, and committed to building trust with their audience. By applying the best practices outlined here and learning from the case studies of those who've gone before you, you can carve out your own profitable path in affiliate marketing.

Closing for Part II

Affiliate marketing is often the **gateway to online income**. It teaches you marketing, content creation, and audience building—all without the burden of creating your own product.

With these skills, you can scale into bigger opportunities like digital products, coaching, or full-fledged e-commerce. But for now, focus on **mastering one niche, one audience, and one promotion channel** until you see results.

Next, in **Part III – Blogging for Income**, we'll explore how to build your own online platform—one of the most powerful vehicles for affiliate marketing and beyond.

Key Takeaways from Part II:

- Affiliate marketing = connecting people with products they already want.
- Pick high-quality products that solve real problems.
- Build trust through content, email, and transparency.
- Success is about **consistency, not shortcuts**.

Part III – Blogging for Income

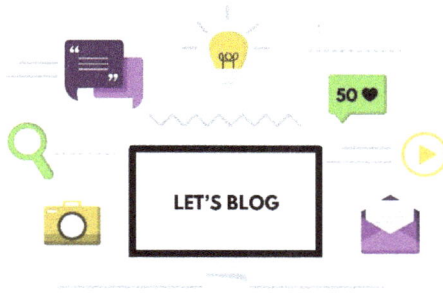

Chapter 1: Setting Up Your Blog

Starting a blog is more than just writing online; it is the foundation of your digital presence. Before content, traffic, and monetization come into play, you must choose the right platform and set up the technical aspects that will make your blog sustainable long-term. From deciding between hosted and self-hosted options (like WordPress.org vs. Wix or Squarespace) to selecting a reliable web host, domain name, and theme, these early decisions determine your blog's scalability and professionalism.

Many beginners underestimate this stage, only to face limitations later. This chapter provides a step-by-step approach to setting up your blog, ensuring it's both functional and future-ready. Think of it as building the house before you decorate and invite guests in.

Before you can make money blogging, you need a strong foundation. That means choosing the right **niche, platform, and design**.

- **Choose a Niche**: The most profitable blogs are often in evergreen niches like **health & fitness, personal finance, relationships, tech, and self-improvement**. But the secret is picking a niche you can stay consistent with for years.
  Example: A nurse who loves food blogging might niche down into *healthy recipes for busy professionals*.
- **Select a Blogging Platform**:
  - *WordPress.org* (recommended for full control and monetization flexibility).
  - *Wix, Squarespace, or Medium* (easy but with limitations).
- **Domain & Hosting**: Invest in a **professional domain name** and a reliable host (e.g., Bluehost, SiteGround, Hostinger).
- **Design & Branding**: Your blog should have:
  - A clean, mobile-friendly theme.

- o Easy navigation (categories, search bar, about/contact page).
- o A strong brand voice and logo.

Checklist – Blog Setup

- • Niche chosen with audience in mind
- • Domain & hosting purchased
- • WordPress (or alternative) installed
- • Professional theme applied
- • About, Contact, and Privacy Policy pages ready

Chapter 2: Content Creation & Overcoming Writer's Block

Content is the heartbeat of blogging. Without valuable, engaging, and purposeful content, a blog is just another empty website floating online. However, not all content is created equal. This chapter focuses on crafting content that not only informs and entertains but also converts—whether that means gaining email subscribers, driving product sales, or attracting brand collaborations.

We will cover frameworks for writing persuasive posts, using storytelling to connect with readers, structuring long-form content for search engines, and balancing evergreen vs. trending topics. Readers will learn how to create blog posts that speak directly to their audience's needs and inspire them to take action.

Great content is the fuel of your blogging business.

- **Types of Blog Content**:
    - *How-to guides* (e.g., "How to Save $500 a Month")
    - *List posts* (e.g., "10 Tools Every Freelancer Needs")
    - *Case studies* (e.g., "How I Grew My Blog to 10,000 Readers")
    - *Opinion posts* (e.g., "Why Affiliate Marketing is the Future of E-commerce")
- **Overcoming Writer's Block**:
    1. Use prompts (*AnswerThePublic, Quora, Reddit*).
    2. Repurpose old posts into new formats.
    3. Use the "Pomodoro technique" (write in 25-min sprints).
    4. Keep an *idea bank* of future post topics.

Checklist – Content Plan

- At least 10 "pillar" blog posts drafted
- 20–30 smaller supporting posts planned
- Editorial calendar created
- Idea bank maintained

Chapter 3: SEO & Backlinks

A blog without traffic is like a shop hidden in the desert—nobody knows it exists. Search Engine Optimization (SEO) ensures that your blog is discoverable, visible, and consistently attracting the right audience. This chapter introduces on-page SEO basics (keywords, headings, internal linking), off-page SEO (backlinks, guest posting, digital PR), and technical SEO (site speed, mobile optimization).

We will explore how to research keywords that match both user intent and profitability, as well as backlink strategies that build authority in your niche. By the end of this chapter, readers will see SEO not as a technical burden but as an organic growth engine for long-term blogging success.

Search Engine Optimization (SEO) ensures your blog gets found on Google.

- **On-Page SEO**:
    - Use **keyword research tools** (Google Keyword Planner, Ubersuggest).
    - Place keywords naturally in titles, meta descriptions, headings, and URLs.
    - Optimize images with *alt text*.
- **Off-Page SEO** (Backlinks):
    - Guest post on bigger blogs in your niche.
    - Build relationships with other bloggers (comment, share their work).
    - Use HARO (Help a Reporter Out) to get featured in media.

Checklist – SEO Basics

- Every post optimized with keywords
- At least 3 internal links & 2 external links per post
- Backlink-building strategy in place

- Blog connected to Google Analytics & Search Console

Chapter 4: Driving Traffic & Outreach

Traffic numbers alone don't guarantee success. What matters is building an engaged and loyal audience. This chapter dives into strategies for transforming casual readers into raving fans who come back repeatedly, share your content, and trust your recommendations. We will discuss building an email list, creating interactive content, leveraging social media to amplify reach, and nurturing reader relationships through comments, community groups, and newsletters.

The emphasis here is on building trust and authority because in the blogging world, credibility equals influence—and influence equals income.

Traffic is the lifeblood of monetization. Without readers, your blog is just an online diary.

- **Free Traffic Sources**:
  - SEO (long-term but powerful).
  - Pinterest (excellent for lifestyle, food, DIY, travel).
  - Social media (Twitter, LinkedIn, TikTok).
  - Quora (answer questions and link back).
- **Outreach Strategy**:
  - Comment on blogs in your niche.
  - Network in Facebook groups & LinkedIn communities.
  - Run collaborations (guest posts, interviews, roundups).

Checklist – Traffic Strategy

- 3–5 social media platforms chosen
- Weekly outreach (guest posts, collabs)
- Pinterest or SEO strategy active

- Monthly analytics review

Chapter 5: Monetization Strategies

At some point, every blogger asks: *"How do I actually make money from this?"* This chapter provides a roadmap for monetization. We'll cover multiple income streams: display ads (Google AdSense, Mediavine, etc.), affiliate marketing (recommending products for commission), sponsored posts (working with brands), and creating/selling your own digital products (courses, ebooks, templates).

Each method has pros, cons, and income potential, and the best bloggers often combine several streams to create stability. Readers will discover how to match monetization methods with their niche and audience size, while also avoiding mistakes like overloading blogs with ads or promoting irrelevant products.

Here is where blogging becomes a business:

- **Affiliate Marketing**: Promote tools/products with affiliate links.
  Example: A fitness blogger promoting workout gear.
- **Ad Revenue**: Use Google AdSense or premium ad networks like Mediavine.
- **Sponsored Posts**: Brands pay you to write about them.
- **Digital Products**: Ebooks, templates, courses, or memberships.
- **Services**: Coaching, freelancing, or consulting.

Checklist – Monetization Prep

- At least 1 monetization strategy chosen
- Affiliate links added to 3–5 blog posts
- Google AdSense or alternative applied
- Plan for first digital product drafted

Chapter 6: Marketing Your Blog

Blogging is a long-term game, and consistency is the silent factor that separates thriving bloggers from quitters. However, every writer eventually hits roadblocks—whether it's running out of ideas, feeling uninspired, or struggling to balance content creation with life's demands.

This chapter equips readers with proven strategies to overcome writer's block, generate endless content ideas, and create a sustainable writing routine. From batching content and using editorial calendars to leveraging AI tools for brainstorming, we will cover practical ways to stay productive without burning out. Because at the end of the day, the blogs that win are not the ones with the most talent but the ones with the most persistence.

A blog without marketing is invisible.

- **Email Marketing**: Build an email list from day one. Use *lead magnets* like free ebooks, checklists, or courses.
- **Social Media Promotion**: Schedule posts via tools like Buffer or Tailwind.
- **Networking & Partnerships**: Team up with influencers and brands.

Checklist – Blog Marketing

- Lead magnet created
- Email marketing platform set up (Mailchimp, ConvertKit)
- Social media content calendar running
- Collaborations planned

Chapter 7: Mistakes, Tips & Lifestyle

- **Common Mistakes**:
    - Spreading too thin across platforms.
    - Focusing only on traffic, not conversions.
    - Giving up too soon (blogs often take 6–12 months to see income).
- **Pro Tips**:
    - Batch-create content.
    - Repurpose blog posts into podcasts, videos, or carousels.
    - Invest in good hosting early to avoid crashes.
- **Blogging Lifestyle**:
  Blogging is flexible but requires discipline. The

freedom to **work from anywhere** is real—but only if you treat your blog like a business.

Final Blogger's Lifestyle Checklist

- Weekly writing routine set
- Monthly analytics review habit built
- Monetization consistently tracked
- Blogging aligned with personal lifestyle goals

Part IV – E-commerce Empire

E-commerce is the backbone of today's digital economy. From small boutique stores selling handmade jewelry to global marketplaces moving billions of dollars' worth of products, e-commerce has revolutionized how people shop and how businesses operate. Building an online store today is more accessible than ever, but success requires more than just setting up a website—you need strategy, branding, and a sharp understanding of your customers.

Chapter 1: Choosing the Right Products

The digital revolution has transformed the way people buy and sell, and e-commerce stands at the center of this transformation. This chapter introduces you to the world of e-commerce, explaining its explosive growth, the opportunities it creates for entrepreneurs, and why it has become one of the most profitable online business ventures in today's global marketplace. Before diving into the practical steps of building an online store, you'll gain clarity on what e-commerce truly means, the different forms it takes, and the reasons it remains a game-changer for both small startups and global giants.

Steps to Find Winning Products

1. **Identify Pain Points** – People buy solutions, not products. For example, ergonomic office chairs address back pain for remote workers.
2. **Follow Trends** – Use tools like Google Trends, TikTok, or Pinterest to identify trending niches.
3. **Validate Demand** – Check Amazon Bestsellers, eBay's trending list, or Etsy's popular products.
4. **Calculate Profit Margins** – Factor in production, shipping, and marketing costs.
5. **Consider Branding Potential** – Can the product stand out with a unique design, packaging, or brand story?

Case Study: Gymshark

Started in 2012 by a teenager in the UK, Gymshark initially sold gym apparel printed at home. By targeting a niche (fitness enthusiasts) and leveraging influencer marketing, it grew into a billion-dollar company.

Checklist: Product Selection

- Solves a problem or meets a strong desire
- Has growing or steady demand
- Healthy margins (ideally 30%+)
- Can be branded or differentiated
- Fits your personal interest/knowledge (optional but useful)

Chapter 2: Building Your Online Store

Every successful business begins with a solid foundation, and in e-commerce, that foundation is your business model. This chapter explores the core models that drive online stores—from retail and wholesale to subscription-based services and marketplaces. You'll discover the strengths and weaknesses of each, and more importantly, learn how to choose the model that aligns with your vision, resources, and long-term goals. By the end, you'll not only understand how these models operate but also be ready to position your store in the right lane for success.

You don't need to be a coder to create a professional e-commerce website. Platforms like **Shopify, WooCommerce, Wix, and BigCommerce** allow anyone to launch an online store with ease.

Key Elements of a Storefront

- **Homepage:** Clean design, value proposition, featured products.
- **Product Pages:** High-quality photos, detailed descriptions, customer reviews.
- **Cart & Checkout:** Simple, fast, and optimized for mobile.
- **Trust Signals:** SSL certificates, clear return policies, and customer testimonials.

Example:

Allbirds (the eco-friendly shoe company) uses storytelling to connect with its customers. Their product pages highlight sustainability, comfort, and quality—convincing buyers that they're purchasing more than just shoes.

Checklist: Online Store Setup

- Mobile-responsive design
- Clear product categories
- Secure payment gateways (Stripe, PayPal, local options)
- Simple checkout process
- "About Us" page with a human story

Chapter 3: Branding & Storytelling

Having the right model is just the beginning; now it's time to build your store. This chapter walks you through the technical and creative steps of setting up a professional online store—from selecting a domain name and hosting plan to designing a customer-friendly interface and integrating secure payment gateways. You'll also learn about compliance requirements, trust-building elements, and how to test your store for a seamless shopping experience. Think of this chapter as your blueprint for turning an idea into a live, functional, and attractive digital storefront.

In e-commerce, branding is the difference between selling a commodity and building a loyal following.

Branding Essentials

1. **Logo & Visual Identity** – Simple, memorable, aligned with your niche.
2. **Brand Story** – Why does your business exist? Who do you serve?
3. **Voice & Tone** – Friendly, authoritative, luxury, or playful—consistency builds trust.
4. **Customer Experience** – Every touchpoint (website, emails, packaging) reflects your brand.

Case Study: Warby Parker

Instead of just selling glasses, Warby Parker built a brand around **affordable fashion eyewear with a social mission** ("Buy a pair, give a pair"). Their brand story resonated deeply with customers, propelling them into a billion-dollar valuation.

Checklist: Branding

- Clear mission & vision
- Unique value proposition (UVP)
- Consistent brand voice
- Strong visual identity (colors, fonts, packaging)
- Emotional connection with customers

Chapter 4: Marketing & Driving Traffic

Not every product is worth selling online, and one of the biggest mistakes new entrepreneurs make is choosing items without proper research. This chapter focuses on the science and psychology behind selecting winning products. You'll learn how to identify problems people are desperate to solve, spot trends before they go mainstream, and balance evergreen products with seasonal hits. By mastering product selection, you'll position yourself ahead of competitors and ensure that every item in your store has the potential to generate sales and build loyalty.

The "build it and they will come" myth doesn't apply to e-commerce. You must drive targeted traffic.

Marketing Channels

- **Paid Ads:** Facebook, Instagram, Google Shopping.
- **Content Marketing:** Blogs, YouTube tutorials, lifestyle posts.
- **Influencer Marketing:** Partnering with niche influencers for authentic promotions.
- **Email Marketing:** Abandoned cart emails, product launches, loyalty campaigns.
- **SEO:** Ranking product pages for keywords like "best ergonomic chair."

Case Study: Beardbrand

Starting as a blog and YouTube channel about beard care, Beardbrand built authority and an audience **before** launching products. Today, they're a leading men's grooming company.

Checklist: Marketing Essentials

- Social media presence (choose 2–3 platforms only)
- Content strategy (blogs, videos, guides)
- Retargeting ads
- Customer reviews and testimonials
- Email sequences (welcome, nurture, re-engagement)

Chapter 5: Scaling & Automation

Once sales start coming in, scaling is the key to building an "empire."

An online store without visitors is like a shop hidden in the desert—nobody knows it exists. This chapter introduces you to the art and science of e-commerce marketing. From search engine optimization (SEO) and social media campaigns to influencer partnerships, email marketing, and paid advertising, you'll discover multiple strategies to drive targeted traffic. Beyond attracting customers, this chapter also emphasizes the importance of building relationships, creating irresistible offers, and establishing trust so that first-time buyers become lifelong customers.

Scaling Strategies

1. **Outsource & Delegate** – Hire VAs for customer service, freelancers for design, agencies for ads.
2. **Optimize Supply Chain** – Bulk orders, better suppliers, faster logistics.
3. **Expand Product Line** – Upsells, cross-sells, and bundles.
4. **International Expansion** – Translate your store, enable global shipping.
5. **Automate Systems** – Inventory, email marketing, analytics, customer support.

Case Study: Kylie Cosmetics

Kylie Jenner leveraged social media influence and Shopify's infrastructure to scale rapidly. With minimal staff, the brand hit **$420M in sales in its first 18 months**, thanks to outsourced manufacturing and distribution.

Checklist: Scaling Your Store

- Reliable suppliers
- Automated inventory and order tracking
- Dedicated customer service channels
- Systems for handling returns/refunds
- Data-driven decision making (Google Analytics, Shopify reports)

Chapter 6: Mistakes to Avoid

Once your store begins generating consistent sales, the next step is growth and freedom. This chapter covers how to scale your business beyond survival mode and into expansion. You'll explore strategies like outsourcing, automation tools, upselling, and global reach to multiply your revenue while reducing your workload. The goal here is to help you build an e-commerce empire that doesn't tie you down but instead runs efficiently—even while you sleep. By the end, you'll see how to transition from being a hustling store owner to a visionary CEO.

Many e-commerce ventures fail because of avoidable errors.

- **Selling "Me-Too" Products** – Don't just copy Amazon bestsellers without differentiation.
- **Neglecting Customer Experience** – Slow shipping, poor communication, no returns policy = lost customers.
- **Relying on One Channel** – If Facebook ads die tomorrow, do you still have traffic?
- **Ignoring Analytics** – Without tracking data, you're flying blind.

Quick Fix Checklist

- Offer multiple payment/shipping options
- Respond to customer inquiries within 24 hrs
- Always test ads with small budgets before scaling
- Monitor store analytics weekly

Wrap-Up: Building Your E-commerce Empire

An e-commerce empire doesn't happen overnight. It's built by:

- Choosing the right products
- Creating a brand that resonates
- Driving consistent, targeted traffic
- Delivering excellent customer experience
- Scaling through automation and smart systems

Whether you dream of running a boutique store from your laptop or scaling into a household name, the blueprint is the

same:  find the

right

product-market fit, tell a compelling story, and deliver value relentlessly.

Part V: Dropshipping

Chapter 1: What is Dropshipping?

Dropshipping has become one of the most popular business models in the e-commerce world. It is often described as a beginner-friendly, low-risk way to start an online business, but the truth is far more nuanced. At its core, dropshipping allows entrepreneurs to sell products online without ever handling inventory or managing warehouses. Instead, store owners act as intermediaries between customers and suppliers. When a customer places an order, the store owner forwards the order to a third-party supplier, who then fulfills and ships it directly to the customer.

The appeal of dropshipping lies in its simplicity and low barrier to entry. Unlike traditional retail, where business owners need to purchase products in bulk and maintain stock, dropshipping eliminates upfront inventory costs. This allows entrepreneurs to focus on marketing, customer acquisition, and brand-building rather than logistics and storage. However, as with any business model, there are misconceptions that need to be cleared. Many people are drawn in by the idea of "easy money" but overlook the challenges—such as low profit margins, supplier issues, and fierce competition.

This chapter aims to **break down the concept of dropshipping in simple terms**, helping you understand not just what it is, but also how it works in practice, and why it has become a global phenomenon. By the end of this chapter, you will have a strong foundation on the basics of dropshipping, allowing you to approach the following chapters with clarity and realistic expectations.

What is Dropshipping?

Dropshipping is a **retail fulfillment method** where a business doesn't keep the products it sells in stock. Instead, when the store sells a product, it purchases the item from a third party (usually a wholesaler, manufacturer, or supplier) and has it shipped directly to the customer.

In simple terms:

1. The customer places an order on your online store.
2. You forward the order to your supplier.
3. The supplier ships the product directly to your customer.
4. You earn the difference between what the customer paid and what you paid the supplier.

Unlike traditional commerce, you don't see, touch, or store the product at any point in the process. This makes dropshipping highly scalable and flexible.

Key Features of Dropshipping

- **No inventory management** – You don't need to invest in or store physical stock.
- **Low startup costs** – You can start with minimal capital compared to brick-and-mortar or traditional e-commerce businesses.
- **Wide product variety** – Since you're not holding stock, you can offer a broad range of products.
- **Location independence** – You can run a dropshipping business from anywhere with internet access.
- **Third-party fulfillment** – Suppliers handle packaging, shipping, and delivery.

Why is Dropshipping Popular Today?

The rise of e-commerce platforms such as **Shopify, WooCommerce, and TikTok Shop** has lowered the barriers to starting a dropshipping store. Combined with the reach of digital advertising (Facebook Ads, TikTok Ads, Google Ads), entrepreneurs can quickly test and scale products without committing to large upfront investments.

The model is especially appealing for:

- **Beginners** who want to test online business with minimal risk.
- **Marketers** who want to focus on advertising and customer engagement rather than logistics.
- **Side hustlers** looking for flexible income opportunities.

Real-Life Example

Imagine you start an online store selling phone accessories. You list a phone case on your website for $20. A customer places an order. You forward the order details to your supplier, who charges you $7 and ships the case directly to the customer. You pocket the $13 difference (minus transaction fees and marketing costs).

You never touched the product, never packed it, and never visited a post office—yet you completed a successful sale.

Conclusion

Dropshipping is not just a buzzword—it is a legitimate business model with both opportunities and challenges. It democratizes e-commerce, allowing anyone with an internet connection and determination to start a business. However, to succeed, one must understand the fundamentals, stay ahead of competition, and build a long-term strategy rather than chasing "get rich quick" myths.

In the next chapter, we will break down the **Dropshipping Business Model Structure** using illustrations and examples to help you see how all the moving parts connect together.

Chapter 2: The Dropshipping Model Structure (Illustrated)

In order to succeed in dropshipping, you need to clearly understand how the business model works at its core. Many new entrepreneurs dive in with excitement but quickly become overwhelmed because they don't have a clear picture of the moving parts. The dropshipping model is simple in theory, but like a well-oiled machine, it only works effectively when all the pieces fit together.

In this chapter, we will break down the structure of dropshipping into clear, practical steps. You'll see how customers, store owners, and suppliers interact in the process. We will illustrate how money, products, and information flow between these parties. By the end, you'll have a mental "blueprint" of the business model—one you can always refer back to as you set up and run your store. This clarity is crucial, because without it, you'll likely waste time, lose money, and struggle to grow.

Understanding the Basic Flow

At its simplest, dropshipping is a **triangular relationship** between three key parties:

1. **The Customer** – The person buying the product from your online store.
2. **The Store Owner (You)** – The entrepreneur running the dropshipping business.
3. **The Supplier/Manufacturer** – The person or company who actually produces, stores, and ships the product.

Here's how the model works step by step:

1. **The Customer Places an Order**
   - The customer visits your online store (Shopify, WooCommerce, etc.), sees a product, and makes a purchase.
   - The money goes to you first.
2. **The Store Owner Transfers the Order to the Supplier**

- After receiving payment, you forward the order details (product + customer shipping information) to your supplier.
- You pay the supplier at the wholesale price.

3. **The Supplier Ships the Product to the Customer**
   - The supplier packages and ships the product directly to your customer under your store's name.
   - You never physically touch or store the product.

4. **Profit Is Made**
   - Your profit is the difference between the customer's purchase price (retail) and the supplier's cost (wholesale).

Illustrating the Model

You can think of the dropshipping model as a **flowchart**:

- **Customer** → Places an order on your **Store**
- **You (Store Owner)** → Forward order & payment (minus profit) to **Supplier**
- **Supplier** → Ships product directly to **Customer**

So the physical product **bypasses you** completely, while the **money flow** goes through you first.

The Role of Technology

This structure is made possible by **technology**. Without modern e-commerce platforms and supplier integrations, dropshipping would be messy and slow. Today, automation tools allow orders to be transferred instantly, tracking numbers to update automatically, and customers to stay informed about their shipments.

Some common tools that power this structure include:

- **Shopify & WooCommerce** for setting up your store.
- **Oberlo, DSers, Spocket, or CJ Dropshipping** for connecting suppliers.
- **Payment processors** (PayPal, Stripe, Paystack) for handling money flow.

These tools ensure the structure runs smoothly, even if you're operating with just a laptop and internet connection.

Key Advantages of This Structure

- **Low Overhead:** You don't need a warehouse or upfront inventory.
- **Flexibility:** You can sell products from multiple suppliers at once.
- **Scalability:** Since suppliers handle shipping, you can scale quickly without storage problems.

Potential Weak Points in the Structure

While simple, this structure does have weak spots:

- **Dependency on Suppliers:** If your supplier is slow, your customer blames you.
- **Thin Profit Margins:** Since you don't buy in bulk, margins are often smaller.
- **Lack of Control:** Packaging, shipping times, and stock availability are outside your control.

This means you must choose reliable suppliers and set clear expectations with customers to keep the model sustainable.

Conclusion

The dropshipping model structure is what makes this business model both appealing and challenging. On one hand, it allows anyone with internet access to run an e-commerce business without the traditional costs of inventory and warehousing. On the other hand, it requires skill in managing relationships with suppliers, keeping track of the customer experience, and ensuring the business flow remains seamless.

With this structure firmly understood, you are now prepared to explore the benefits and challenges of dropshipping, which we'll cover in the next chapters.

Chapter 3: Advantages of Dropshipping

O ne of the key reasons dropshipping has gained massive popularity is its accessibility. Traditional retail and e-commerce require significant capital investment, inventory management, and logistical operations. Dropshipping, however, flips this model on its head by minimizing risk and lowering the entry barrier for aspiring entrepreneurs.

This chapter explores the major **advantages of dropshipping**, explaining why it has become one of the most appealing online business models. From the **low startup cost** to **flexibility, scalability, and location independence**, dropshipping creates opportunities for both beginners and experienced entrepreneurs.

By understanding these benefits, you'll have a clearer perspective on why dropshipping could be the right choice for you and how to use these advantages strategically to maximize profit while minimizing risk.

The Advantages of Dropshipping

1. Low Startup Costs

One of the biggest advantages of dropshipping is that you don't need to invest heavily upfront. Unlike traditional retail models where you buy stock in bulk, store it, and then sell, dropshipping requires **no inventory investment**. Your primary costs are:

- A domain name and hosting (for your website or store).
- Marketing and advertising campaigns.

- Optional tools for automation (like apps, analytics, or customer service).

This makes dropshipping an affordable entry point into e-commerce, especially for entrepreneurs who cannot risk thousands of dollars upfront.

2. No Need for Warehousing

Storing products, paying for warehouses, and managing logistics can quickly eat into profits in traditional businesses. With dropshipping:

- Your supplier handles warehousing.
- You save on rent, utility bills, and security costs.
- There's no need for inventory management systems.

This significantly reduces operational stress and allows you to focus solely on sales, marketing, and customer acquisition.

3. Wide Range of Products to Sell

Because you're not tied down by physical stock, you can test and sell virtually **any product category** without restrictions. This means:

- You can easily pivot to new niches if one isn't profitable.
- You can offer thousands of products without additional costs.
- You can continuously update your store with trending or seasonal items.

This flexibility gives you an edge in adapting to market demands quickly.

4. Location Independence

Dropshipping allows you to run your business **from anywhere in the world**. Whether you're at home, traveling, or living abroad, all you need is a laptop and internet connection. This freedom makes dropshipping particularly attractive to digital nomads and people seeking lifestyle businesses.

5. Scalability

In traditional retail, growth means more staff, larger warehouses, and increased overheads. Dropshipping, on the other hand, scales efficiently:

- Adding new products doesn't require storage expansion.
- Suppliers handle increased order volume.
- Automation tools can manage customer support, orders, and marketing at scale.

This scalability allows small businesses to grow into large brands without massive infrastructure.

6. Reduced Risk

Since you don't pre-purchase inventory, there's **little to no risk** of being stuck with unsold products. This means:

- You won't face losses from overstocking.
- You can test products with minimal investment.
- You can stop selling a failing product instantly without worrying about liquidating inventory.

This makes dropshipping a **low-risk business model**, especially for beginners.

7. Easy to Get Started

With platforms like **Shopify, WooCommerce, or TikTok Shop**, setting up your online store can take just a few hours. You don't need technical expertise, coding knowledge, or years of retail experience. Modern tools and templates simplify the process.

8. Product Testing Made Simple

For entrepreneurs who love experimenting, dropshipping is perfect. You can test multiple products and niches at the same time without committing to bulk purchases. This rapid testing capability helps you identify **winning products** faster than traditional models.

Key Takeaway

Dropshipping eliminates many of the barriers that typically hold people back from starting a business. Its **low cost, flexibility, and scalability** make it one of the most attractive e-commerce models in today's digital age.

However, while these advantages are impressive, they do not come without challenges. In the next chapter, we'll balance the perspective by looking at the **disadvantages of dropshipping**—the hidden pitfalls and risks you must prepare for.

Chapter 4: Disadvantages of Dropshipping

While dropshipping has opened doors for countless entrepreneurs to step into the e-commerce space without heavy upfront investments, it is not without challenges. Every business model has trade-offs, and dropshipping is no exception. Many newcomers rush into the model because of the low entry barriers, but fail to account for the operational, financial, and reputational risks that can quickly drain profits or ruin a brand.

Understanding these disadvantages does not mean abandoning dropshipping altogether. Instead, it prepares you to make informed decisions, implement preventive measures, and build a business with realistic expectations. This chapter explores the most common drawbacks of dropshipping, from slim profit margins to supplier-related risks, and highlights strategies to mitigate them.

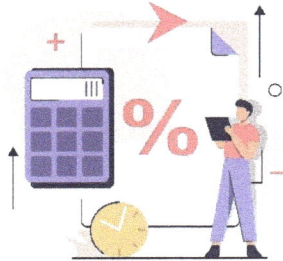

The Major Disadvantages of Dropshipping

1. Low Profit Margins

- **Explanation:** Since dropshipping has low entry costs, the market is flooded with sellers offering similar products. This drives prices down, leaving you with razor-thin margins.
- **Example:** If you sell a product for $25, but your supplier charges $18, and after paying $4 for shipping and $3 for advertising, your profit is close to zero.
- **Mitigation:** Focus on niche markets or unique products where competition is lower, and brand value can justify higher pricing.

2. Heavy Reliance on Suppliers

- **Explanation:** You don't control stock, packaging, or shipping. If a supplier runs out of stock, ships late, or sends defective products, your reputation suffers, not theirs.
- **Example:** A customer orders a Christmas gift from your store, but the supplier fails to deliver on time. The customer blames you, not the supplier.
- **Mitigation:** Work with multiple suppliers, test their reliability, and keep open communication.

3. Longer Shipping Times

- **Explanation:** Many dropshipping suppliers, especially from overseas (e.g., China), have shipping times ranging from 2 to 6 weeks. In an era of Amazon Prime and instant gratification, this frustrates customers.
- **Mitigation:** Use local suppliers where possible or partner with fulfillment services (e.g., 3PL warehouses, or programs like CJ Dropshipping that offer faster shipping options).

4. Limited Branding and Customization

- **Explanation:** Since you don't own or control product manufacturing, it's hard to customize packaging, add personal touches, or differentiate your product. This weakens brand identity.
- **Mitigation:** Consider private labeling once your business scales or negotiate branded packaging with suppliers.

5. High Competition

- **Explanation:** With low startup costs, anyone can become a dropshipper. This saturates markets with look-alike stores. Customers often buy from whoever has the cheapest price or fastest delivery.
- **Mitigation:** Build a brand story, offer excellent customer service, and use content marketing to stand out.

6. Customer Service Challenges

- **Explanation:** When things go wrong (delayed shipping, damaged products, wrong item sent), you are caught in the middle. You don't handle the inventory directly, yet customers expect you to fix problems immediately.
- **Mitigation:** Create clear policies, communicate transparently, and work with suppliers who respond quickly to issues.

7. Refunds and Returns Hassles

- **Explanation:** Handling returns can be complicated. Suppliers may not accept returns, or the cost of sending items back overseas might be higher than the product's value.
- **Mitigation:** Build a clear refund policy and, where necessary, refund customers directly without asking them to return items if it's not worth the logistics.

8. Less Control Over Quality

- **Explanation:** You don't inspect items before they reach customers. Low-quality or inconsistent products can lead to bad reviews and lost trust.
- **Mitigation:** Order test products yourself, and read supplier reviews before partnering with them.

9. Advertising Dependency

- **Explanation:** Since dropshipping doesn't usually rely on organic brand loyalty at first, most sales come from paid ads (Facebook, TikTok, Google). Advertising costs can eat up profits quickly.
- **Mitigation:** Diversify traffic sources — use SEO, influencer marketing, and email marketing to reduce reliance on paid ads.

10. Legal and Trademark Risks

- **Explanation:** Some suppliers may sell counterfeit or copyrighted products without your knowledge. Selling these can lead to lawsuits or store bans.
- **Mitigation:** Avoid branded items (e.g., Nike shoes, Apple products). Stick to generic products or partner with legitimate wholesalers.

Summary

Dropshipping is often promoted as a "get rich quick" model, but the reality is that it comes with significant challenges. From slim margins and stiff competition to customer service headaches and lack of control over supply chains, these disadvantages can break a business if unaddressed. However, acknowledging these risks allows you to plan ahead, build resilience, and transition from just another dropshipping store to a **sustainable, branded e-commerce business**.

Chapter 5: E-Commerce Platforms for Dropshipping

O ne of the most important decisions in your dropshipping journey is choosing the right **e-commerce platform**. Think of your platform as the foundation of your business—it's where your store lives, how your products are showcased, and how customers interact with your brand. While the dropshipping model frees you from the burden of managing inventory or warehousing, your platform will determine how seamless your operations are, how professional your store looks, and how easily you can scale.

The truth is, there's no "one-size-fits-all" solution. Some platforms are perfect for beginners, others are better for scaling brands, and a few are ideal if you're highly tech-savvy. In this chapter, we'll break down the most popular platforms for dropshipping, analyzing their strengths, weaknesses, and suitability depending on your goals. By the end, you'll be able to identify the platform that fits your needs and avoid wasting time on the wrong setup.

Key E-Commerce Platforms for Dropshipping

1. Shopify

- **Overview**: Shopify is the most popular platform for dropshipping. It's beginner-friendly, integrates with thousands of apps, and offers a sleek, customizable storefront.
- **Pros**:
  - Easy to set up (no coding required).
  - Wide range of dropshipping apps like Oberlo, DSers, Spocket.
  - Professional themes and user-friendly dashboard.
  - Scalable for both small stores and large brands.

- **Cons**:
    - Monthly subscription fee ($29+).
    - Transaction fees if not using Shopify Payments.
    - Some essential apps cost extra.
- **Best for**: Beginners and intermediate dropshippers who want a professional and scalable store.

2. WooCommerce (with WordPress)

- **Overview**: WooCommerce is an open-source plugin that transforms a WordPress website into an online store. It's flexible, customizable, and widely used.

- **Pros**:
  - Free to install (you only pay for hosting, domain, and add-ons).
  - Full control over your website.
  - Countless plugins for dropshipping (AliDropship, WooDropship, etc.).
  - Good for content + store integration (great for blogging and SEO).
- **Cons**:
  - Requires more technical skills to set up and maintain.
  - Can get complicated as you add more plugins.
  - Updates and security issues must be managed manually.
- **Best for**: Entrepreneurs who want more control and have some technical skills or a budget for a developer.

BIGCOMMERCE

3. BigCommerce

- **Overview**: A hosted e-commerce platform like Shopify but with slightly different pricing and features.
- **Pros**:
    - No transaction fees.
    - Built-in marketing tools.
    - Scales well with high-volume stores.
    - Multi-channel integration (Amazon, eBay, Facebook).
- **Cons**:
    - Less beginner-friendly than Shopify.
    - Fewer third-party app integrations compared to Shopify.
- **Best for**: Growing dropshipping businesses with higher budgets that want to sell across multiple channels.

WiX.com

4. Wix eCommerce

- **Overview**: A drag-and-drop website builder that added e-commerce functionality.
- **Pros**:
    - Very easy to use.
    - Affordable pricing compared to Shopify.
    - Nice templates and beginner-friendly design.
- **Cons**:
    - Limited compared to Shopify/WooCommerce for dropshipping integrations.
    - Not as scalable for long-term growth.
- **Best for**: Absolute beginners testing the waters or those who want a simple side business.

5. Squarespace

- **Overview**: Popular among creatives and small businesses. Known for its beautiful templates.
- **Pros**:
  - Aesthetic and professional designs.
  - Easy to build and manage.
- **Cons**:
  - Limited dropshipping plugins.
  - Less functionality compared to Shopify or WooCommerce.
- **Best for**: Small niche dropshipping stores focused on design-conscious audiences (fashion, jewelry, lifestyle).

5. Other Options

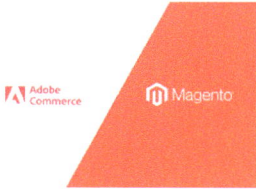

- **Magento (Adobe Commerce)**: Highly customizable but complex. Best for large businesses with developer resources.

- **Etsy (integrated with dropshipping suppliers)**: Works if you're focusing on print-on-demand or handmade-inspired products.

amazon.com

ebay

- **Amazon & eBay (marketplace selling)**: Lower control over branding but massive traffic potential.

Choosing the Right Platform: Key Factors to Consider

1. **Budget** – Can you afford monthly fees like Shopify, or do you prefer the low-cost but higher-effort WooCommerce?
2. **Ease of Use** – Do you want plug-and-play simplicity (Shopify/Wix) or flexibility with a learning curve (WooCommerce)?
3. **Scalability** – Will the platform grow with your business?
4. **Integrations** – Does it support the dropshipping apps and suppliers you want to use?
5. **Design & Branding** – Does the platform allow your store to stand out visually?

Practical Example

- A student with no technical skills, looking to quickly launch a dropshipping store, might choose **Shopify** for its simplicity and integrations.
- A blogger who already has a WordPress website and wants to monetize with dropshipping products might lean toward **WooCommerce**.
- A growing business with ambitions to scale across Amazon, eBay, and multiple stores might invest in **BigCommerce**.

Conclusion

The e-commerce platform you choose is like the foundation of your online store. While Shopify often dominates the conversation, it's not the only option. WooCommerce, BigCommerce, Wix, and others all have unique strengths depending on your goals. The key is to balance **cost, scalability, ease of use, and flexibility** to match your vision for dropshipping success.

Chapter 6: Product and Supplier Sourcing

Product and supplier sourcing is the lifeline of any dropshipping business. Since you don't physically stock or manufacture the products, your suppliers essentially become your business partners. Choosing the right suppliers and sourcing strategies will determine whether your business thrives or collapses. Reliable suppliers ensure smooth order fulfillment, good product quality, and fast shipping — all of which build customer trust.

On the flip side, poor supplier choices can result in bad reviews, refund requests, and a ruined brand reputation. Therefore, learning how to effectively source products and build relationships with trustworthy suppliers is critical to long-term success. This chapter explores where to find reliable suppliers, how to vet them, and strategies to secure the best products at competitive prices.

Understanding Product and Supplier Sourcing

Supplier sourcing is the process of identifying, evaluating, and partnering with companies that will fulfill your customer orders. In dropshipping, suppliers handle inventory storage, packaging, and delivery on your behalf. Therefore, the suppliers you choose directly impact the quality of customer experience.

Product sourcing, on the other hand, involves selecting the items you want to sell. A winning dropshipping business is built on sourcing products that have proven demand and suppliers who can deliver them reliably.

Where to Source Suppliers

1. **Dropshipping Marketplaces**
   - **AliExpress**: The most popular platform for beginner dropshippers due to its wide product variety and no upfront cost.
   - **CJ Dropshipping**: Offers faster shipping times than AliExpress and a wide range of product categories.
   - **Spocket**: Focuses on suppliers in the US and EU, enabling faster delivery to Western markets.
   - **SaleHoo**: A directory of vetted suppliers with wholesale and dropshipping options.
2. **Wholesale Directories & Networks**
   - **Worldwide Brands**: A directory of certified wholesalers and dropshippers.
   - **Doba**: Aggregates multiple suppliers into one platform, simplifying order management.

3. **Local Manufacturers & Suppliers**
   - Partnering with local suppliers ensures faster shipping, easier communication, and higher product quality.
4. **Custom Product Suppliers**
   - Private label suppliers allow you to sell customized or branded products, giving your store a unique edge.

How to Vet Suppliers

Not every supplier is worth working with. You need to evaluate them carefully:

- **Product Quality**: Always order test samples before listing products in your store.
- **Shipping Times**: Suppliers with faster and more reliable shipping options improve customer satisfaction.
- **Communication**: Good suppliers respond quickly to questions and updates.
- **Return Policy**: Ensure they have a clear return/refund system in place.
- **Reviews & Ratings**: Look for suppliers with a strong track record of positive feedback.

Building Strong Supplier Relationships

Dropshipping is more sustainable when you treat your suppliers as partners:

- **Communicate Clearly**: Keep them updated on order volumes, seasonal changes, or special requests.
- **Negotiate Better Deals**: Once you establish consistent sales, negotiate lower prices or exclusive arrangements.
- **Develop Backup Options**: Always have multiple suppliers for your winning products in case one fails.

Strategies for Product Sourcing

1. **Test Multiple Suppliers**: Start by testing different suppliers for the same product to compare shipping, packaging, and quality.

2. **Focus on Niche-Specific Suppliers**: Instead of general marketplaces, look for suppliers specializing in your niche (e.g., fitness gear, beauty products).

3. **Use Data Tools**: Platforms like Jungle Scout, Ecomhunt, or AliShark can help identify trending products and reliable suppliers.

4. **Dropshipping Agents**: Consider hiring an agent in China who sources, checks, and ships products on your behalf.

Common Mistakes in Supplier Sourcing

- Choosing a supplier solely based on low prices (ignoring quality).
- Failing to order test products before selling.
- Not having backup suppliers for winning products.
- Ignoring communication speed and customer support.

Conclusion

Your suppliers are the backbone of your dropshipping store. Great products and fast shipping can build your brand, while unreliable suppliers can destroy it overnight. By carefully sourcing both products and suppliers, testing them, and building professional relationships, you create a stable foundation for your business.

Chapter 7: Winning Product Formula

E very successful dropshipping business thrives on one crucial element: the right product. In a crowded e-commerce market, simply listing random items on your store isn't enough. A winning product is more than just something that sells. It's an item that solves a problem, sparks an emotion, or fulfills a strong desire for a particular audience. It's the product that makes customers stop scrolling, pay attention, and click "buy now."

Understanding the formula for identifying winning products can make the difference between a profitable store and one that struggles. In this chapter, we will break down the attributes of winning products, explore the psychology behind why people buy, and present a structured formula you can follow to consistently identify and sell items that convert.

The Science Behind a Winning Product

A winning product stands at the intersection of **demand, uniqueness, profitability, and emotional appeal**. It's not about selling what you like—it's about selling what the market wants.

Think about it this way: people buy products for one of three main reasons:

1. To solve a problem (pain-relief products).
2. To make life easier or more enjoyable (convenience/lifestyle products).
3. To express themselves or feel better (emotional products).

The best products often tick at least two of these boxes.

The Winning Product Formula

Here is a simple formula you can use to evaluate products before investing time and money into them:

Winning Product = Problem Solving + Wow Factor + High Perceived Value + Healthy Profit Margin + Demand Proof

Let's break each of these down:

1. Problem Solving

- Products that fix a real issue are easier to market.
- Example: A posture corrector solves back pain from sitting too long.
- Ask yourself: Does this product address a pain point or frustration in people's lives?

2. Wow Factor

- Does the product grab attention within 3 seconds of seeing it?
- Products that spark curiosity or excitement tend to go viral on social media.

- Example: A 3-in-1 portable blender is more eye-catching than a regular kitchen blender.

3. High Perceived Value

- Customers buy based on perceived value, not actual cost.
- Example: A luxury-looking watch that costs $8 to source but can retail at $50+ because of branding and presentation.
- Strong branding, sleek packaging, and good product images can elevate perceived value.

4. Healthy Profit Margin

- A good dropshipping product should allow at least **2.5x to 3x markup**.
- If you source a product for $10, you should aim to sell it for $25–$30 or more.
- Remember, marketing eats into your costs—without a margin, you're not running a business.

5. Demand Proof

- Are people already buying this product?
- You can verify demand by checking platforms like TikTok, AliExpress, Amazon, or eBay.
- If you see multiple sellers and growing interest, it's a positive sign.

Additional Attributes of Winning Products

- **Lightweight & Easy to Ship:** Bulky items increase shipping costs and customer complaints.
- **Impulse Buying Potential:** Best dropshipping products don't require deep consideration. Customers buy on emotion, not logic.
- **Low Competition or Untapped Market:** Avoid saturated products unless you can brand them uniquely.
- **Room for Upsells:** Products that allow cross-selling or bundling increase profits. For instance, selling a pet grooming brush alongside pet nail clippers.

Examples of Winning Products

1. **Pet Industry:** Self-cleaning cat litter box (problem-solving + wow factor).
2. **Fitness Industry:** Resistance bands with training app (value add + perceived value).
3. **Beauty Industry:** LED facial mask (trend-driven + high markup).
4. **Kitchen Niche:** Automatic vegetable slicer (time-saving + impulse buy).

Case Study: The Portable Blender

- **Problem Solving:** Allows smoothies on-the-go.
- **Wow Factor:** Compact, rechargeable, visually appealing.
- **High Perceived Value:** Marketed as a lifestyle gadget, not just a blender.
- **Profit Margin:** Sourced at $12, sold at $39.99.
- **Demand Proof:** Millions of TikTok views and consistent Amazon sales.
  → This product checks all the boxes of the formula.

Action Steps for You

1. List 10 products you're considering.
2. Score each on the five factors (Problem-solving, Wow factor, Perceived value, Profit margin, Demand).
3. Narrow down to 2–3 products that score highest.
4. Test ads with small budgets before scaling.

Key Takeaway:

A winning product is not luck—it's strategy. By applying this formula, you'll minimize risk, increase your chances of success, and position yourself ahead of other dropshippers who simply pick products at random.

Chapter 8: Where to Find Winning Product Ideas

One of the most exciting yet challenging parts of dropshipping is discovering *winning products*. A winning product is not just anything you stumble upon; it's a product that solves a problem, excites customers, and generates consistent sales. Many beginners mistakenly think success in dropshipping comes from flashy ads or fancy websites. The truth is, even the best marketing won't save you if the product doesn't appeal to buyers.

This chapter will walk you through proven strategies and tools to uncover products that have the potential to go viral or dominate in a specific niche. You'll learn how to analyze trends, monitor competitors, and use data-driven research to separate potential winners from products that will waste your time and resources. By the end, you will know exactly where and how to search for your next winning product.

Finding Winning Product Ideas

1. **Social Media Platforms**
   Social media is one of the best hunting grounds for dropshippers because it is where trends are born and spread quickly. Platforms like TikTok, Instagram, and Facebook showcase what people are currently excited about.
   - o **TikTok:** Search hashtags like *#TikTokMadeMeBuyIt* or *#ViralProducts* to discover trending items with high engagement.
   - o **Instagram & Pinterest:** Visual platforms that give insight into lifestyle trends, gadgets, and fashion.
   - o **Facebook Ads Library:** Browse running ads to see what products other dropshippers are pushing successfully.

2. **E-commerce Marketplaces**

   Large online marketplaces can give clues about what's selling well globally.

   - **Amazon Best Sellers:** A goldmine for identifying high-demand categories and trending items.
   - **AliExpress Hot Products:** Shows items gaining traction with dropshippers.
   - **Etsy:** Especially good for unique, creative, or personalized items.
   - **eBay Trending:** Useful for spotting products with growing buyer demand.

3. **Competitor Research Tools**

   Instead of reinventing the wheel, learn from what's already working.

   - **Dropshipping Spy Tools** like AdSpy, Sell The Trend, and Ecomhunt help you track winning products and their ad performance.
   - Analyze competitor stores to see which products they are consistently promoting.

4. **Trend Analysis Platforms**

   Use tools that analyze search trends and predict rising demand.

   - o **Google Trends:** Shows search popularity of products over time.
   - o **Trend Hunter:** Highlights new trends in fashion, lifestyle, and tech.
   - o **Exploding Topics:** Focuses on up-and-coming product categories before they go mainstream.

5. **Customer Review Mining**

   Sometimes the best ideas come from simply listening to customers.

   - o Read product reviews on Amazon or AliExpress to see what people like or dislike.
   - o Pay attention to recurring complaints—these often highlight opportunities for improved or alternative products.

6. **Offline Inspiration**

   Winning products aren't always found online.

   o Visit local malls or popular retail stores like Walmart or Target to see trending gadgets and seasonal products.

   o Observe what products are heavily promoted in physical stores; often, they perform well online too.

Tips for Evaluating Product Potential

When you find a product idea, ask yourself:

- Does it solve a problem or fulfill a need?
- Is it unique or hard to find locally?
- Is it affordable yet profitable?
- Does it spark an emotional reaction (wow factor)?
- Can it be marketed with compelling visuals and videos?

Conclusion

Finding winning products isn't about luck; it's about observation, research, and timing. By combining social media trend-spotting, marketplace insights, competitor analysis, and trend tools, you can build a steady pipeline of profitable product ideas. Remember: the product is the *foundation* of your dropshipping business—get this right, and everything else becomes easier.

Chapter 9: Factors to Consider When Choosing the Right Niche

One of the most important decisions you'll make as a dropshipper is **choosing your niche**. Think of your niche as the "territory" where your business will play — it's the specific category of products you'll sell and the audience you'll serve. Picking the wrong niche can lead to frustration, low sales, and wasted money, while the right niche can open doors to consistent profits and long-term growth.

Many beginners make the mistake of picking a niche simply because it's trending or because they saw someone else succeed with it. But success in dropshipping doesn't just come from copying — it comes from **strategically analyzing and selecting a niche** that has demand, profitability, and aligns with your personal interests.

In this chapter, we will explore the **key factors you should consider when choosing your niche**, from profitability and demand to competition and product quality. By the end, you'll understand how to critically evaluate a niche before committing your time and resources to it.

Key Factors to Consider When Choosing a Niche

1. Market Demand

- Look for niches where customers are already spending money.
- Tools like **Google Trends**, **Amazon Bestsellers**, **eBay Trending**, and **TikTok hashtags** can help you see if people are actively searching for and buying products in that niche.
- A good niche should have a balance: consistent demand, but not so trendy that it fades quickly.

Example:
Fitness products like resistance bands or yoga mats show

consistent demand, while fidget spinners were a short-lived trend.

2. Profitability

- Dropshipping often works with thin margins, so you must ensure the niche supports products with **healthy profit margins**.
- A good profit margin usually falls between **20–50%** after considering product cost, shipping, ads, and platform fees.
- Avoid niches where products are cheap but require heavy marketing spend to sell.

Example:

Phone cases may cost $2–$3 but require high ad costs to sell for $10, leaving slim profits. On the other hand, ergonomic office chairs may sell for $150+ with a good margin even after expenses.

3. Competition Level

- If a niche is overcrowded with big brands, it may be hard to stand out.
- Use tools like **SEMRush**, **Ahrefs**, or even a simple search on Google/Amazon to see how saturated the niche is.
- Don't be scared of competition; instead, look for niches where you can bring a **unique twist** — for example, targeting eco-friendly yoga mats instead of generic yoga mats.

4. Target Audience

- Understand **who your buyers are**. A niche is only as strong as the audience behind it.
- Ask yourself:
  - Who buys these products?
  - Are they impulse buyers or careful planners?
  - Do they hang out on platforms like TikTok, Instagram, or Pinterest?
  - Do they have disposable income?

- Niches that attract passionate communities (pet lovers, gamers, fitness enthusiasts) are usually strong because buyers are emotionally connected to the products.

5. Personal Interest and Knowledge

- You don't have to be a hardcore fan of your niche, but having **some level of interest or knowledge** helps.
- When you understand your products and audience, it's easier to create convincing marketing campaigns, write product descriptions, and build trust.
- Many successful dropshippers choose niches aligned with their hobbies or lifestyle.

Example:
A gamer who dropships LED gaming accessories will likely create more authentic marketing than someone with no idea about gaming culture.

6. Product Quality and Reliability

- Even if a niche seems profitable, poor product quality will ruin your business reputation.
- Check supplier reviews and customer feedback on product quality, delivery times, and consistency.
- Remember, in dropshipping, **your brand carries the blame** for poor-quality products, not the supplier.

7. Seasonality

- Some niches are **seasonal**, meaning they sell only during certain times of the year (e.g., Christmas decorations, Halloween costumes).
- Seasonal niches can be profitable, but you should balance them with **evergreen niches** that sell all year long.
- Google Trends can help identify if interest in a niche spikes only during certain months.

8. Legal and Compliance Issues

- Avoid niches that may cause legal problems.
- Restricted or high-risk products (like weapons, medical equipment, counterfeit items, branded knock-offs) can get your store banned.
- Always ensure your products comply with e-commerce platform policies and local laws.

9. Scalability

- A good niche isn't just profitable now — it has the potential to grow.
- Ask yourself:
    - Can I expand my product range within this niche?
    - Can I upsell or cross-sell related products?
    - Will this niche still be relevant in 2–3 years?
- A scalable niche gives you room to grow into a brand rather than being stuck with one-hit products.

Practical Example: Comparing Two Niches

- **Niche 1: Pet Accessories**
  - Market Demand: Strong and consistent (pet owners spend heavily on pets).
  - Profitability: Good margins on items like automatic feeders, pet beds, or GPS collars.
  - Audience: Passionate pet lovers.
  - Scalability: Easy to expand (toys, clothing, health items).

    Strong niche.

- **Niche 2: Fidget Toys**
  - Market Demand: Trendy but short-lived.
  - Profitability: Small margins, high ad costs.
  - Audience: Mostly kids/teens, not much disposable income.
  - Scalability: Limited.

    Weak niche.

Action Steps for Choosing the Right Niche

1. Brainstorm a list of 5–10 potential niches.
2. Use Google Trends to check demand consistency.
3. Research competition using Amazon, TikTok, and e-commerce platforms.
4. Calculate potential profit margins.
5. Narrow down to 2–3 niches and test products with small ad campaigns.

Conclusion:

Choosing the right niche is one of the most critical steps in dropshipping. It determines your audience, product selection, marketing approach, and ultimately, your profitability. By carefully analyzing factors like demand, competition, profitability, and scalability, you set yourself up for long-term success rather than chasing short-lived trends.

Chapter 10: Common Mistakes to Avoid in Dropshipping

Every business model has pitfalls, and dropshipping is no exception. Many entrepreneurs rush into it believing it's a quick way to make money online without understanding the intricacies involved. While dropshipping has low entry barriers, it requires patience, attention to detail, and a strategic mindset. Unfortunately, countless beginners repeat the same mistakes, leading to wasted money, customer dissatisfaction, and eventual business failure.

This chapter aims to highlight the most common mistakes in dropshipping, why they happen, and—most importantly—how to avoid them. By understanding these mistakes, you will be better positioned to build a sustainable and profitable dropshipping business.

1. Choosing the Wrong Niche

- **The Mistake:** Many new dropshippers chase "trending" products or pick niches they don't understand. This often leads to low demand, poor customer interest, or high competition.
- **How to Avoid:** Conduct thorough niche research (see Chapter 9). Focus on niches with demand, profitability, and long-term growth. Choose products that solve problems or appeal to emotions.

2. Relying on a Single Supplier

- **The Mistake:** Depending on just one supplier leaves your business vulnerable to stock shortages, shipping delays, or supplier issues.
- **How to Avoid:** Always have backup suppliers for your products. Build relationships with multiple reliable suppliers to ensure business continuity.

3. Ignoring Product Quality

- **The Mistake:** Many dropshippers prioritize cheap suppliers without checking the actual quality of products. This leads to high return rates, negative reviews, and loss of customer trust.
- **How to Avoid:** Order test samples before selling. Read supplier reviews, and choose suppliers with proven track records for product quality.

4. Long Shipping Times

- **The Mistake:** Relying on suppliers with extremely long delivery times (sometimes 30–60 days) results in frustrated customers.
- **How to Avoid:** Partner with suppliers offering faster shipping options like ePacket, local warehouses, or third-party logistics providers. Be transparent with customers about shipping timelines.

5. Poor Customer Service

- **The Mistake:** Many dropshippers neglect customer inquiries or fail to resolve complaints quickly, assuming that because they don't handle stock, service doesn't matter.
- **How to Avoid:** Prioritize communication. Respond to queries promptly and with empathy. Offer refunds or replacements where necessary. Excellent customer service sets you apart from competitors.

6. Overpricing or Underpricing Products

- **The Mistake:** Beginners often price products either too high (scaring off customers) or too low (leaving little to no profit margin).
- **How to Avoid:** Calculate all costs, including product cost, shipping, transaction fees, and marketing expenses. Then set a reasonable profit margin while remaining competitive.

7. Lack of Marketing Strategy

- **The Mistake:** Thinking products will sell themselves. Many entrepreneurs build a store but fail to invest in effective marketing.
- **How to Avoid:** Use social media ads (Facebook, Instagram, TikTok), influencer marketing, SEO, and email campaigns. Focus on brand storytelling and targeted advertising to reach your ideal audience.

8. Poor Website Design

- **The Mistake:** Launching a store with a poorly designed, slow, or unprofessional website undermines credibility. Customers may leave before making a purchase.
- **How to Avoid:** Invest in a clean, user-friendly design. Ensure your store is mobile-friendly, loads quickly, and has clear product descriptions and images.

9. Ignoring Legal and Business Formalities

- **The Mistake:** Many dropshippers overlook tax requirements, refunds, privacy policies, and terms & conditions. This can cause legal issues down the line.
- **How to Avoid:** Research the legal requirements of your country and your customers' countries. Draft clear policies for refunds, shipping, and privacy. Register your business if required.

10. Giving Up Too Early

- **The Mistake:** Dropshipping is often portrayed as an overnight success business. Many beginners quit after a few weeks if they don't see quick profits.
- **How to Avoid:** Understand that dropshipping is a long-term business. It takes time to test products, optimize ads, and build trust. Persistence and consistent learning are key.

Conclusion

Avoiding these common mistakes doesn't guarantee instant success, but it does protect you from unnecessary setbacks. Dropshipping requires effort, research, and adaptability. By sidestepping these pitfalls and focusing on customer satisfaction, efficient operations, and smart marketing, you will significantly increase your chances of building a profitable and sustainable dropshipping business.

Chapter 11: What is TikTok Shop and Benefits of TikTok Shop

In recent years, social media has transformed from being just a platform for entertainment into a powerful driver of e-commerce. TikTok, with its massive user base and highly engaging short-form video format, has gone a step further by integrating a marketplace directly within its app: the **TikTok Shop**. For dropshippers and e-commerce entrepreneurs, this represents a golden opportunity to not only reach but also convert audiences where they are most engaged.

TikTok Shop allows sellers to list products directly on the platform, enabling users to make purchases seamlessly without leaving the app. With over 1 billion active users globally, TikTok offers not just visibility but also a strong conversion channel driven by **authentic content, algorithm-driven reach, and influencer collaborations**. This chapter explores what TikTok Shop is, how it works, and why it has become a powerful tool for dropshippers looking to maximize sales.

TikTok Shop

What is TikTok Shop?

TikTok Shop is TikTok's built-in e-commerce feature that enables brands, creators, and merchants to sell products directly within the app. It integrates with videos, live streams, and a dedicated shop tab, creating a **frictionless shopping experience**.

Unlike traditional e-commerce platforms, TikTok Shop merges **entertainment and shopping** into a single space; giving rise to the term "shoppertainment." With its powerful algorithm, TikTok shows products to people who are most likely to purchase, based on their viewing habits and interactions.

How TikTok Shop Works:

- Sellers upload and list their products within TikTok Shop.
- Products can be featured in **videos, live streams, and the shop tab.**
- Viewers click the product link and purchase **without leaving TikTok.**
- TikTok handles the payment system and ensures smooth order processing.

Benefits of TikTok Shop for Dropshippers

TikTok Shop is particularly attractive to dropshippers because it aligns perfectly with their model of low upfront costs, high engagement, and reliance on trending products. Here are the main benefits:

1. Access to a Massive Audience

TikTok has a global reach with billions of users across different demographics. For dropshippers, this is a goldmine of potential customers. The platform is especially popular with Gen Z and millennials, groups known for their **high purchasing power in online shopping**.

2. Algorithm-Driven Visibility

TikTok's recommendation system pushes content to people who are most likely to engage. This means that even new sellers without large followings can achieve viral reach if their content resonates with the audience.

3. Seamless Shopping Experience

Unlike traditional ads that redirect customers to an external site, TikTok Shop allows users to buy directly within the app. This **reduces friction** in the buying process and significantly increases conversion rates.

4. Low-Cost Marketing through Content

Instead of spending heavily on paid ads, dropshippers can create short, engaging videos showcasing their products. With creativity, even small sellers can compete with big brands by going viral organically.

5. Influencer & Affiliate Opportunities

TikTok Shop has a built-in affiliate system where influencers and creators can promote your products in exchange for a commission. This allows dropshippers to leverage TikTok influencers without upfront marketing costs.

6. Trend-Driven Sales

TikTok thrives on trends — from viral challenges to trending sounds. Dropshippers can ride these waves by pairing trending content with their products, creating a **massive sales boost in a short time.**

7. Lower Entry Barriers

Unlike platforms that require significant advertising spend to gain traction (like Facebook or Google Ads), TikTok Shop allows sellers to start small and scale quickly based on product performance.

Key Features of TikTok Shop

- **Shoppable Videos:** Sellers can tag products in regular TikTok videos.
- **Live Shopping:** Sellers can host live streams, demo their products, and sell instantly.
- **Affiliate Program:** Creators earn commissions for promoting products.
- **Shop Tab:** A dedicated tab on TikTok profiles where all listed products can be viewed.
- **Analytics Dashboard:** Tools to track sales, traffic, and conversions.

Why TikTok Shop is a Game-Changer for Dropshipping

TikTok Shop isn't just another platform; it's reshaping how dropshipping works. Traditional methods required **driving traffic from social media ads to e-commerce stores**, but TikTok removes that step, making the process more direct and efficient.

Dropshippers now have the advantage of:

- Selling directly on a high-engagement platform.
- Leveraging viral content for **zero-cost customer acquisition.**
- Building trust faster through relatable influencer collaborations.

In other words, TikTok Shop gives dropshippers a **built-in marketplace powered by social proof and entertainment.**

Conclusion

TikTok Shop is more than just an e-commerce feature. It's a powerful blend of entertainment and shopping, designed for today's social media-driven consumers. For dropshippers, it opens up opportunities for **viral growth, influencer partnerships, and simplified selling.** By tapping into TikTok Shop, dropshippers can stay ahead of the curve in an ever-changing e-commerce landscape.

Conclusion

Your Roadmap to Online Business Mastery

You've now journeyed through the many stages of building, scaling, and mastering an online business—from laying the foundation, learning income streams, mastering marketing, creating content that sells, to cultivating the mindset of an entrepreneur. By this point, you should recognize that online business is not just about making money; it is about creating freedom, impact, and a lifestyle that aligns with your values.

But mastery doesn't happen overnight. Just as a skilled artisan hones their craft through repetition, failure, and persistence, the online entrepreneur builds mastery through consistent learning, experimentation, and resilience.

Here is a **simplified roadmap** you can always refer back to:

1. **Start Small, Start Smart** – Choose one niche, one platform, and one monetization method to begin with. Don't spread yourself too thin.
2. **Create Value** – Focus relentlessly on solving problems and adding value to people's lives. The money follows the value.
3. **Build Systems, Not Just Income** – Automate, delegate, and streamline so your business doesn't rely entirely on you.
4. **Keep Learning & Adapting** – The online landscape evolves quickly; staying static means falling behind.
5. **Think Long-Term** – Short-term wins are exciting, but sustainable growth and mastery come from patience and vision.

This roadmap is not a one-way street. You'll revisit steps, re-learn lessons, and pivot when necessary. But as long as you keep moving forward, you're already ahead of those who only dream but never take action.

Call to Action: Start Today

Knowledge without execution is just potential energy—it only becomes power when you act on it. Many aspiring entrepreneurs get caught up in endless planning, overthinking, and waiting for the "perfect time." The truth? There is no perfect time. The best time to start was yesterday; the next best time is now.

Here's your challenge: **Choose one action from this book and commit to doing it within the next 24 hours.**

- Register your domain.
- Write your first blog post.
- Record that video.
- Post your offer online.
- Reach out to a potential partner.

Take one step. Then another. Then another.

Remember, mastery is not built on giant leaps—it's built on daily, consistent actions that compound over time.

As you close this book, I want you to imagine yourself one year from today. If you apply what you've learned, take action consistently, and stay committed, you could have a profitable online business, freedom to design your lifestyle, and impact thousands of lives with your work.

So don't just dream. Don't just plan. **Start today.**

Your journey to online business mastery begins with one step—take it now.

Appendix

Helpful Materials & Resources

To support your journey as an online entrepreneur, here are carefully selected tools, platforms, and resources:

1. E-commerce Platforms

- Shopify – www.shopify.com
- WooCommerce – www.woocommerce.com
- BigCommerce – www.bigcommerce.com
- Wix eCommerce – www.wix.com/ecommerce

2. Product & Supplier Sourcing

- AliExpress – www.aliexpress.com
- Alibaba – www.alibaba.com
- CJ Dropshipping – www.cjdropshipping.com
- Spocket – www.spocket.co

3. Marketing & Advertising Tools

- Facebook Ads Manager – www.facebook.com/business/tools/ads-manager
- TikTok Ads Manager – www.tiktok.com/business
- Google Ads – ads.google.com
- Canva (for design) – www.canva.com

4. Research & Winning Products

- Ecomhunt – www.ecomhunt.com
- Sell The Trend – www.sellthetrend.com
- Dropship Spy – www.dropshipspy.com
- TrendHunter – www.trendhunter.com

5. Learning & Skill Building

- HubSpot Academy (Free Courses) – academy.hubspot.com
- Coursera – www.coursera.org
- Udemy – www.udemy.com
- Oberlo Blog – www.oberlo.com/blog

6. Community & Support

- Reddit: r/dropship
- Shopify Community – community.shopify.com
- Facebook Groups (Search "Dropshipping & E-commerce")

***Note:** The author has no direct affiliation with the above companies or platforms. These are suggested solely as helpful resources for your growth.*

www.ingramcontent.com/pod-product-compliance
Lightning Source LLC
Chambersburg PA
CBHW040852210326
41597CB00029B/4814